RECKONING IN COAL COUNTRY

By Mason Adams
and Dustin Bleizeffer

Written by Mason Adams and Dustin Bleizeffer
Edited and designed by Lani Hanson and Ken Paulman
Original series edited by Katie Klingsporn and Ken Paulman
Cover photo by Dustin Bleizeffer

Mason Adams has worked as a journalist since 2001, covering Appalachian communities and the issues that affect them. Born and raised in Clifton Forge, Virginia, Mason previously worked as a wildlife biologist before moving into journalism. He lives in Floyd County, Virginia.

Dustin Bleizeffer has worked as a coal miner, an oilfield mechanic, and for 22 years as a statewide reporter and editor primarily covering the energy industry in Wyoming. He was a John S. Knight Journalism Fellow at Stanford and also served as WyoFile's editor-in-chief. He lives in Casper.

TABLE OF CONTENTS

INTRODUCTION

The coronavirus pandemic brought the demise of U.S. coal into sharp focus in 2020. After a decade of decline marked by a series of bankruptcies among coal giants and smaller operators alike, the economic shutdown from the pandemic revealed coal's growing disadvantage in power markets while demonstrating the monumental task of economic and cultural transition that lies ahead for coal communities.

The implications are far reaching, but perhaps nowhere are the stakes higher than in Central Appalachia and Wyoming — the engines of the American coal industry.

WyoFile and Energy News Network, with support from the Just Transition Fund, teamed up to produce the "Transition in Coal Country" series, tapping two journalists with local knowledge and experience covering the industry; Mason Adams in Virginia and Dustin Bleizeffer in Wyoming.

The six-part series examines what's driving coal's irreversible decline, how it's playing out in the heart of coal country, and what's at stake for communities as they envision paths forward. Despite their geographic, cultural and socioeconomic differences, Wyoming and Central Appalachia together exemplify what's at stake in the midst of a massive energy transition; the nation cannot afford to leave vast regions of coal country behind.

These are regions that "literally fueled the growth of the nation," said Peter Hille, president of the community economic development nonprofit Mountain Association in eastern Kentucky.

"There is a debt to be paid. Justice demands we bring new investment to these places: to build a new economy, to revitalize communities and to educate people of all ages to be ready."

Neither region is without hope or their own homegrown ingenuity to face an uncertain future. But time is not on their side.

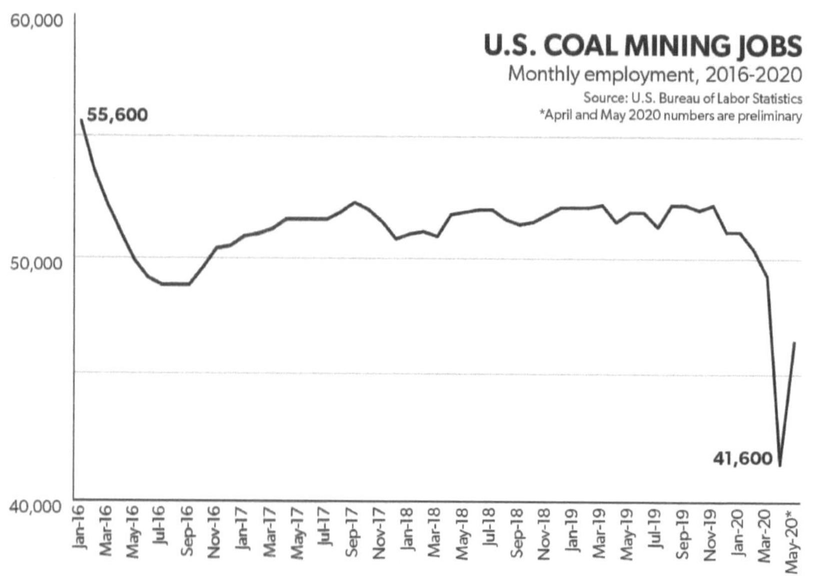

U.S. COAL MINING JOBS
Monthly employment, 2016-2020
Source: U.S. Bureau of Labor Statistics
*April and May 2020 numbers are preliminary
55,600
41,600
60,000
50,000
40,000
Jan-16
Mar-16
May-16
Jul-16
Sep-16
Nov-16
Jan-17
Mar-17
May-17
Jul-17
Sep-17
Nov-17
Jan-18
Mar-18
May-18
Jul-18
Sep-18
Nov-18
Jan-19
Mar-19
May-19
Jul-19
Sep-19
Nov-19
Jan-20
Mar-20
May-20*

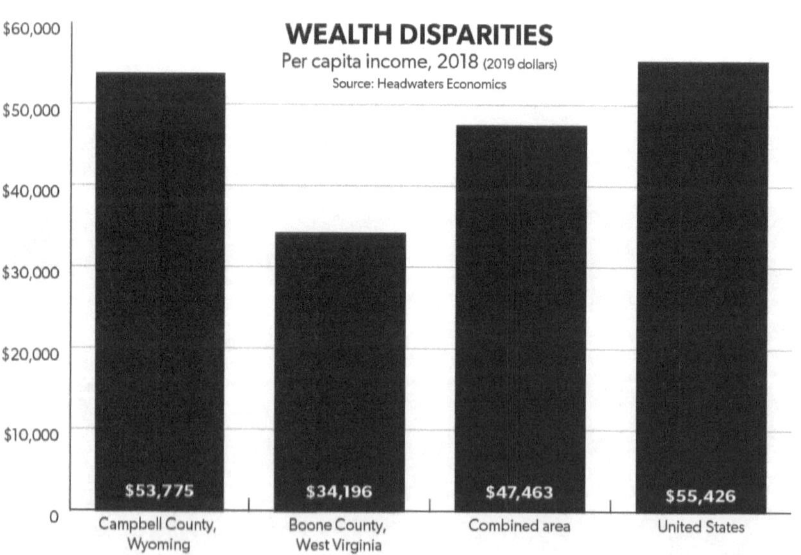

WEALTH DISPARITIES
Per capita income, 2018 (2019 dollars)
Source: Headwaters Economics
$60,000
$50,000
$40,000
$30,000
$20,000
$10,000
0
$53,775
$34,196
$47,463
$55,426
Campbell County, Wyoming
Boone County, West Virginia
Combined area
United States

CHAPTER ONE

WHAT'S NEXT FOR COAL COUNTRY?

Coal communities in Appalachia and Wyoming are vital revenue engines for huge portions of the rural U.S., and they face devastation with little time to adjust to new realities.

Coal country in the United States is in the midst of a historic transition. Whether it's Welch, West Virginia; or Cumberland, Kentucky; or Gillette, Wyoming, communities that have built their cultures and economies around coal face a future without it. This transition is not one most coal communities would have chosen, and because of the economic shock of the 2020 coronavirus pandemic, it's transpired much sooner than anticipated.

The decline that many coal communities expected to stretch years into the future has instead arrived, seemingly, overnight. Coal's demise was already baked into market and policy forces driving toward cleaner energy for the past 20 years, according to analysts. But the coronavirus pandemic's economic implications abruptly exposed the industry's weaknesses to hasten its downfall. Rather than a "glide path" of a decade or more to diversify from the economic base of coal, communities in coal country may see that foundation erode in just a few years.

"Basically the [Wyoming revenue] trend that's happened here is a vertical-downward; there's no slope, it's just straight down," University of Wyoming energy economist Robert Godby said. "It's like the elevator is not just plunging down the elevator shaft; the cable broke and we're just going straight down."

By 2016, coal production in both Wyoming and West Virginia

had already dropped by half from 2008. Since then, the decline has only accelerated — driven by competition from cheap natural gas, tougher restrictions on pollution, and the declining cost of solar and wind energy.

Wyoming's coal mining industry lost 20% of its customer base in the past 10 years and will lose another 23% over the next 10 years as the nation's fleet of coal-fired power plants continues to shrink, according to a report by the University of Wyoming and Montana-based Headwaters Economics. The same trend applies to coalfields in Appalachia, where only mines that produce metallurgic coal anticipate continuing markets in steelmaking.

Coal-fired power plants and coal mines have served as job and revenue drivers for regions that stretch far beyond a coal town's borders. Mineral extraction from just a handful of counties accounts for the bulk of Wyoming's revenue: 52.2% in 2017, down from 67.6% in 2006. Downturns in coal and oil in Wyoming are primary drivers behind an estimated $1.5 billion revenue loss from 2021 through 2022 — an amount equal to half the state's non-education side of the biennium budget.

Coal's decline brings significant climate and public health benefits. But without an effective transition to find new ways to sustain these communities, entire swaths of the rural U.S. face economic devastation.

WYOMING AND APPALACHIA

The Powder River Basin and central Appalachia remain the largest producers of coal in the U.S., with Wyoming and West Virginia as the No. 1 and No. 2 producing states. Together, Wyoming and Appalachia produce two-thirds of the nation's coal, and they're the two areas that will be hit hardest by the industry's decline.

They're also two very different regions, with significant distinctions when it comes to demographics, socioeconomics and the coal industry itself.

Vast wealth disparities exist between Wyoming and Appalachian coal country. For example, personal income in Campbell County, Wyoming, grew by more than 717% from 1970 to 2018 compared to 47% in Boone County, West Virginia, according to federal data compiled by Headwaters Economics. By 2018, per capita income in Campbell County was $53,775 compared to $34,196 in Boone County.

Neither region is monolithic. Eight states make up the Appalachian coal region, each a multitude of cultural, political and economic identities. Wyoming too is made up of distinct cultures and economies, from the blue-collar oil and coal town of Gillette on the eastern high plains to the tony destination resort town of Jackson on the outskirts of Teton and Yellowstone national parks.

Together, Wyoming and Appalachia tell the story of United States coal communities in the midst of certain change with uncertain outcomes. Their successes and failures will help shape the future of rural life in the U.S.

Yet despite the industry's outsized influence, neither region is destined to be permanently ruined by the loss of coal.

Wyoming, for one, is not destitute. The state has generally managed to retain much more wealth than its coal counterparts in the eastern U.S., mostly because mining is largely concentrated on federal coal holdings that have returned billions of dollars to Wyoming coffers. The state of just 578,000 people has built up more than $28 billion in savings and investments.

Although it may never make up for the scale of revenue and salaries lost in coal, Wyoming is only just beginning to tap into a growing outdoor recreation industry. More than half of Wyoming's landscape is made up of public lands, home to abundant wildlife and opportunities to recreate and isolate in wide open vistas. Business leaders also hope to draw on the state's robust industrial services sector and skilled labor force to attract more manufacturing and tech jobs.

Similarly, it would be limiting to describe Appalachia as just "coal country." Of the 13 states included in the Appalachian Regional Commission's definition of the region, only eight produce coal: West Virginia, Pennsylvania, Kentucky, Alabama, Virginia, Ohio, Mississippi and Tennessee, in order from most coal production in 2018 to least, according to data from the U.S. Energy Information Administration. The central Appalachian coal region has seen significant losses in coal production due to the expense of getting at coal seams, which tend to be deeper and thinner than in the Powder River Basin. Coal's sharp post-2012 drop-off has deeply affected central Appalachia's coal communities, which tend to be rural and highly dependent on coal and related industries such as the railroad and equipment manufacturing. A 2018 study of the coal industry for the Appalachian Regional Commission found major decreases in the prime working-age population in mining counties, while the poverty rate has increased.

Hope exists in Appalachia, though, with a burgeoning outdoor recreation industry spurring growth among restaurants and other tourism-related businesses. Some communities have used music, history and traditional crafts to attract visitors as well. Others are investing in broadband in hopes of drawing tech jobs and remote workers, or in retraining miners for the clean energy sector.

Chris Woolery, a staffer at Kentucky's Mountain Association, a nonprofit looking for new ways to create jobs, said there's no "silver bullet" for replacing coal. Instead, communities should try using a lot of "silver BBs" instead.

NEW CHAOS THREATENS TRANSITION EFFORTS

One of the most challenging aspects of coal's demise is that coal communities have little to no control in how quickly it unfolds. Coal country's officials and residents often say they imagined the end of coal remained far off in the future, and that as it wound down, mines would be safely reclaimed and sealed, offering communities time to prepare for the future.

But bankruptcy laws and feeble regulation enforcement to ensure full reclamation of mining's impacts have erased the illusion of control over an orderly conclusion to coal's role as the United States' energy driver.

Waves of bankruptcy have riddled the industry as companies have used reorganization and liquidation to dump mines that produce little coal but carry millions of dollars' worth of environmental and employee liabilities. The chaos generated by these bankruptcies effectively hangs miners and their communities out to dry, and potentially leaves taxpayers to foot the bill.

Take Blackjewel, whose bankruptcy attracted national attention when a group of laid-off miners who'd not been paid for their final weeks of work blocked a coal-laden train from leaving a mine near Cumberland, Kentucky. Blackjewel was part of a spate of bankruptcies in recent years by coal companies whose holdings consisted largely of mines they had obtained through previous bankruptcies by larger companies such as Alpha Natural Resources, Arch and Peabody.

Blackjewel's bankruptcy led to a scramble for its remaining productive mines, particularly the Eagle Butte and Belle Ayr mines — Alpha's former "crown jewels" in the Powder River Basin. Potential bidders simultaneously attempted to avoid getting stuck with the company's more unproductive mines in central Appalachia, which produce little coal but come with tens of millions of dollars each in reclamation work that remains to be done. Advocacy group Appalachian Voices estimated that more than 60 mining permits could wind up abandoned, leaving the clean-up job to the states.

The Blackjewel bankruptcy dragged on through 2019 and 2020, with creditors ranging from vendors to state and local governments trying to get paid. Blackjewel's lawyers have accused former CEO Jeff Hoops of looting the coal company through various family businesses.

None of this bodes well for coal or the communities that produce it.

"Blackjewel may be unfortunately the future of the coal industry," said Clark Derry-Williams, an energy finance analyst at the Institute for Energy Economics and Financial Analysis. "Essentially, you had a company take over assets from other companies' bankruptcies, mismanage them and turn them into an operational and financial mess as well. That's what I fear is going to happen to more and more of the coal industry in the U.S."

That was the state of the coal industry even before COVID-19 arrived. The virus was slow to reach and spread through coal country, but the economic shutdown in response to the pandemic triggered a recession and further slowed the industry. Meantime, regional watchdog groups say it's critical to hold coal companies accountable to employee obligations and mine reclamation in order to soften the economic blow to mining communities.

"We have watched this play out in Appalachia to the detriment of that region's waters, lands and workers, and we have tried to apply those lessons to our work here in Wyoming," said Shannon Anderson, attorney for the Wyoming landowner advocacy group Powder River Basin Resource Council. "One thing Appalachia can learn from us is that as the industry declines, it is the very worst time to weaken standards, reduce bonding and reclamation obligations, and otherwise give companies a free pass."

TRANSFERABLE LESSONS, AND PATHS FORWARD

Although much of Appalachia shares a common history around coal, it's a region where the industry exists in multiple stages simultaneously. Some Appalachian communities ceased mining in the 1950s and remember their coal heritage only in historical markers, while others are only just beginning to accept that coal will never be the economic driver it once was.

In that sense, Appalachia has been struggling with the transition from coal for more than 70 years. Its communities have already been forced to experiment with how to survive beyond a coal-dominated economy.

Those experiments range wildly: They include the outdoor adventure economy that's taken root in Fayetteville, West Virginia, and tech research in Pittsburgh's Innovation District. Dozens of small towns market their roots music heritage and folk art. In eastern Kentucky, U.S. Rep. Hal Rogers talks up "Silicon Holler," while across the state line in Virginia, a public-private economic development project launched in 2019 is targeting "future of work"-style jobs in energy innovation, advanced manufacturing and other fields.

These initiatives have met mixed success. One lesson to emerge is there's no one-size-fits-all answer for coal communities. While the opportunities for transition hinge on a similar set of truths — a community's ability to evolve depends upon good schools, healthcare services, and amenities such as elder care and cultural resources — much depends on what a community actually wants, and whether it has the leadership to help build that vision.

Responding to the inevitable losses in coal jobs and revenue, Colorado established a "Just Transition" collaboration and a fund to help coal communities reorient toward more sustainable economies. A similar strategy was launched in New Mexico to divest from coal-fired power while providing financial aid to communities to prepare for the transition away from coal.

But there's still no comparable effort underway in Wyoming, where the political response to coal's calamity, so far, is to fight against the forces driving it. In 2019, the Wyoming Legislature passed a bill requiring utilities planning to retire coal-fired units ahead of schedule to first offer them for sale to a third party. The next year, the Legislature passed a bill requiring utilities in the state to install financially risky carbon capture systems to coal-fired units.

For Gillette and Campbell County in Wyoming, the most agreeable first steps to diversify lie in a long ambition to build an industry of coal-to-products research and manufacturing. Although it will never replace the scale of jobs and revenue that might be lost in mining, many locals still hold out hope that it will help slow the retirement rate of coal-fired plants among thermal coal's customer base in the U.S., while creating new markets for exporting coal technology and coal-derived products.

State and local economic development officials envision their own version of Silicon Valley in northeast Wyoming, which they've coined "Carbon Valley."

"Campbell County is an energy community," said Phil Christopherson, CEO of Energy Capital Economic Development. "We want to continue to develop that, and include and expand renewables."

In 2019, the Wyoming Legislature granted $5 million of a $10 million request from Gov. Mark Gordon for a pilot project using advanced technologies to capture at least 75% of carbon emissions from a coal-fired power plant of 5 megawatts or more built in Wyoming.

Others say the state should deliberately shift away from spending money on coal, and instead invest more in efforts to provide assistance to coal workers and their communities.

"We will be better off when our state leaders stop wasting time and money trying to prop up this failing industry and start using the limited resources that we have to plan for and support a controlled transition," said Sierra Club Wyoming Chapter director Connie Wilbert.

Wilbert said she believes that frustration and resentment still guide state policy regarding the decline of coal — a massive economic and cultural force in Wyoming for many decades. But the state risks even more by not taking steps today toward a post-coal future.

"The Sierra Club, absolutely, we have played a powerful role in helping push this change along and helping reveal all of the hidden costs" of coal, Wilbert said. "At the same time, we also absolutely believe that we need to do everything in our power to find ways to help the people who are most impacted by this change find another way forward — actually help bring everyone, all of us, along together so we don't have such extreme winners and losers."

In eastern Kentucky, Carl Shoupe worked in the mines until he was injured, and then became an organizer and activist. In July 2019, on the second day of the Blackjewel train blockade, Shoupe stood in solidarity with the protesting miners. Even as Harlan County residents increasingly accepted the growing reality that coal is dying, Shoupe said the community itself will bear the burden of moving forward.

"As long as the coal companies were producing and making money and donating to these politicians to help elect them, everything was fine," Shoupe said. "But now, see, there's no money going to the politicians in the state — no revenues to amount to anything as far as coal severance tax. For the state, it's a no-brainer; they could give a pop what goes on here. And I honestly believe they'll leave us here with the bag to hold."

Without intervention, coal's decline could leave Wyoming looking more like Appalachia, with lower per-capita income and higher poverty rates. And continuing depopulation could drop Appalachia from its current population density of between 40 and 60 people per square mile to something closer to Wyoming's — six people per square mile.

Yet, there's also room for optimism. The people of both regions are renowned for their independence, tenacity and ability to endure tough circumstances. Coal's end does not also mean the end of the communities it once dominated.

The future of coal country remains unwritten. That story will be determined by those who live there now.

Photo by Dustin Bleizeffer / WyoFile

Like many hospitals across the nation, Campbell County Health imposed furloughs and salary reductions during the economic fallout of the coronavirus pandemic hoping to return employees to regular hours.

CHAPTER TWO

FACING A HEALTHCARE CRISIS

The coronavirus pandemic hammered healthcare systems for rural populations already struggling with the loss of coal jobs and health coverage.

Even before the coronavirus pandemic hit, the only hospital serving a vast swath of Wyoming's coal country was in financial trouble.

Gillette-based Campbell County Health's revenue began to slip after a 2016 downturn in coal, marked by hundreds of lay-offs at Powder River Basin mines in northeast Wyoming. The sprawling complex of surface mines produces nearly 40% of the nation's coal.

A series of coal company bankruptcies followed, along with more layoffs. Campbell County Health Chief Operating Officer Colleen Heeter said the community began losing healthcare workers because they moved away with spouses who were let go from the mines.

A "ransom" cyber attack hit the hospital in September 2019, arresting its operations for two weeks — costing it an estimated $2 million to $3 million. Then COVID-19 struck, causing patients to put off visits and elective procedures — sapping 50% of the hospital's revenue.

More layoffs and furloughs at the mines followed, on top of hundreds of jobs that suddenly disappeared in Powder River Basin oil fields. Energy analysts warned the worst is yet to come; they predicted some mines in the region might close within the year.

"We know we have to make some changes as it relates to the new normal," Heeter said.

The coal industry's outsized influence on local healthcare plays out across the nation in coal communities, which tend to be located in rural areas where providing a full slate of healthcare services was a struggle long before coal's backslide and the COVID-19 pandemic.

While healthcare providers are struggling in rural communities across the country, the situation at Campbell County Health illustrates the compounding economic crises of the coronavirus pandemic and a historic downturn in coal — forces that strike particularly hard for coal communities in both central Appalachia and Wyoming.

Hospitals in both regions have long struggled with aspects of their rural circumstances: chronic health conditions, aging populations and too few people to pay for the increasing costs of healthcare. Rural coal communities that had enjoyed a patient base with a moderate level of private insurance are seeing more patients rely on Medicaid and Medicare. They're also seeing an alarming increase in uncompensated care.

Nationally, "the average rural hospital has a 4% operating margin, and most of that margin comes from services provided to people with private insurance," said Beth O'Connor, executive director of the Virginia Rural Health Association. "If you take those services away, nothing else is paying the bills. I would be amazed if everybody survives."

There have been 15 closures among rural U.S. hospitals so far this year, according to the University of North Carolina's Rural Health Research Program. The number is likely to grow substantially before year's end.

"Everybody is scrambling," said Debrin Jenkins, executive director of the West Virginia Rural Health Association. "These hospitals were already running in the red even before this."

HOSPITALS AT THE BREAKING POINT

In 2020 alone, three hospitals closed in West Virginia, all in coal-producing counties. Fairmont Regional Medical Center in Marion County and Bluefield Regional Medical Center in Mercer County closed in mid-March and late July, respectively, but the emergency departments of both have subsequently re-opened. Williamson Memorial Hospital in Mingo County closed in April. More hospital closures seem likely to follow.

Hospitals in coal country were structured to rely on a steady tax base from the industry, as well as good-paying jobs that provide health benefits to employees and their families. The permanent losses in coal, along with its effects on ancillary businesses, have destabilized the existing system of how healthcare is paid for and delivered. It's a monumental challenge for rural coal communities, and an essential component of a community's chances of survival beyond coal.

Robust healthcare services are as crucial as housing, roads, water and schools if any rural community is to survive, let alone reinvent itself after the loss of a legacy industry that paid most of the bills. Yet coal's decline has undercut the ability of local businesses and governments to invest in maintaining those assets.

"It's the perfect storm for those communities," said Eric Boley, president of the Wyoming Hospital Association and previously administrator at South Lincoln Medical Center in Kemmerer. "Even prior to COVID-19, the downturn in coal had a really detrimental impact on the communities in a lot of different ways."

For decades, the coal town of Kemmerer, Wyoming, has relied on the Naughton power plant, associated coal mine and a pair of natural gas processing plants to boost local revenue and help financially support healthcare. One of the three coal-burning units at Naughton powered down in 2019; another will go offline in 2025 while the third is converted to natural gas — actions that will cost hundreds of jobs at the power plant and the Kemmerer Mine.

Rock Springs, Glenrock, Gillette and other Wyoming energy towns faced similar circumstances before the COVID-19 pandemic hammered coal and the oil and gas industries.

"Now, you've got hospitals that are literally hemorrhaging [money] because they've stopped all of the [elective] procedures and all of the things that actually bring revenue into their hospitals," Boley said in an April 2020 interview. "COVID-19 has just had a total catastrophic impact on healthcare in our state."

Even as medical providers lean on telehealth technology, many of their patients have poor cell service and worse internet. O'Connor said that one southwestern Virginia hospital still relies on film mammography because it doesn't have the broadband capacity for a digital machine.

At the same time, coal's decline continues to sap revenue for hospitals. For example, a portion of Campbell County Health's funding comes from a mill levy based on property taxes; the healthier the local economy, the more money it generates for the hospital. The county's three mills dedicated to the hospital generated $18 million three years ago — barely enough to cover the hospital's annual uncompensated care, CCH's Heeter said. Now, thanks to the downturn in coal and oil, those same three mills will deliver about $11.2 million in 2020.

That budget-altering challenge is still trending downward.

Some Wyoming hospitals reported losing up to 70% of revenue one month into the pandemic, Boley said. Approximately 1,539 people had lost their jobs in the healthcare and social services sectors in Wyoming from mid-March to the end of April 2020 and were receiving unemployment benefits, according to the Wyoming Department of Workforce Services. That doesn't account for healthcare staff who moved out of the community after their spouses were laid off from coal and oil jobs.

Like many hospitals across the nation, Campbell County Health is desperately trying to stem the loss of healthcare professionals by imposing pay cuts and furloughs before turning to layoffs.

"Things are only going to get worse," Boley said. "If things don't turn around quickly, I have some serious concerns about whether or not a lot of our hospitals will even survive this. And I'm not being dramatic — I'm being realistic when I take a look at the amount of cash they're burning to try to stay open and trying to keep their communities protected. We've got some tough times ahead of us."

By mid-May 2020, Wyoming hospitals and healthcare providers had received two injections of CARES Act relief funds totaling about $139 million, a lifeline that likely prevented one or more hospital closures, Boley said. Uncompensated care is accelerating and hospitals are simply trying to emerge from the pandemic intact.

Many Wyoming hospitals resumed non-emergency procedures after the Centers for Medicare and Medicaid Services revised its COVID-19 guidelines in April 2020. Central Appalachian hospitals resumed in late April and early May. However, hospitals everywhere were still limited in the number of non-emergency procedures they could take on due to continuing protective measures for COVID-19 — and many people continued to put off seeking care from fear they may be exposed to the virus.

REACTING TO A PANDEMIC IN A COAL CRISIS

Congress included $100 billion for hospitals in the $2 trillion CARES Act, which staunched some of the fiscal bleeding. But rural hospital leaders say the relief was just a bandage.

"There's not enough money that Congress can print to underwrite the cost of all of this," said Ballad Health CEO Alan Levine. "The faster we get our operations back up, the faster we get back to financial stability."

Ballad Health is Southwest Virginia's dominant hospital chain. The nonprofit, which also covers northeast Tennessee, formed through the merger of two previously competing nonprofit chains that were both losing millions of dollars. Virginia did not expand Medicaid until 2019, after two hospitals had closed. Tennessee, which still hasn't expanded Medicaid, has seen 13 hospital closures since 2000.

Ballad received $38 million from the CARES Act, but "$38 million doesn't make up for more than $100 million in terms of cash flow that we've lost … from where we were a year ago," Levine said. Additionally, Ballad received a $200 million advance on Medicare that it must pay back.

The hospital chain was expected to manage through the fall of 2020, Levine said, but the future was still hazy because of uncertainty about how the pandemic would play out.

"Only 17% of our in-patient revenue is covered by commercial insurance, 60% Medicare and 15% is Medicaid," Levine said. "If there's a small reduction in the number of people covered by insurance, or an increase in the number of people who have high deductibles, all of that could have a major effect financially on the system. We're watching all of that in terms of our cash flow."

The closure of a clinic or hospital creates an economic ripple that has an outsized effect on rural places that already lack large pools of employers.

"For most rural communities, the hospital is one of the top three employers; usually the school district and hospital are one and two," O'Connor said. "If you take that hospital away, not only are you losing individual jobs, but those are good-paying jobs — doctors and nurses and respiratory therapists and whatnot — good jobs with good benefits paying a lot into the tax base of a community."

Suppliers and other support businesses also take a hit, as well as restaurants and retail establishments that benefit from having doctors, nurses and other medical personnel as customers.

The problems afflicting coal country's hospitals began well before the pandemic, and relief money from the CARES Act and other sources will, at best, keep them going only for the short-term. More comprehensive solutions remain elusive, especially given the political gridlock in Washington, D.C., and the limitations of state funding support.

Meanwhile, healthcare systems were bracing for more waves of coronavirus — whether as states began to reopen, or when cold weather returned.

Photo by Mason Adams

Patients receive care at a free clinic at the Wise County Fairgrounds in 2019.

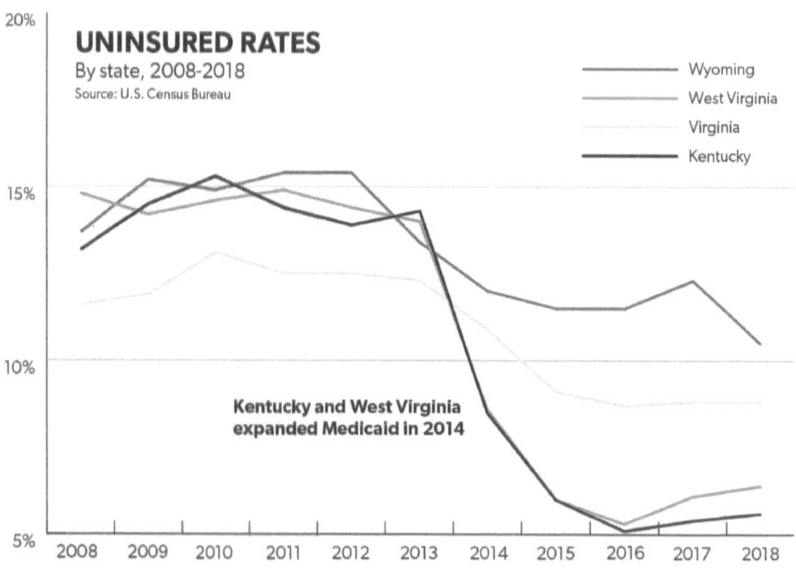

CHAPTER THREE

INCREASING RELIANCE ON FEDERAL HEALTH PROGRAMS

Coal's decline has forced staunchly anti-Obamacare politicians to reconsider Medicaid expansion, while rural communities depend on telehealth more than ever.

Every July for two decades, volunteer doctors, dentists and nurses gathered at the Wise County Fairgrounds to deliver free medical, dental and vision care.

The free clinic attracted thousands of central Appalachian residents, many who camped out the night before to secure their spot in line, as well as media outlets from around the world that were drawn to the dramatic images of people receiving healthcare in trailers and livestock stalls.

But in 2020, the three-day free clinic sponsored by Remote Area Medicine didn't happen. That was due partly to social distancing restrictions in response to the coronavirus pandemic, but also because RAM's local partners "have built infrastructure and capacity around eye, dental, and medical resources in the Wise, Virginia community," according to a news release. Local organizations such as the Health Wagon and Mission of Mercy had grown to fill the gaps and were now able to stage regular health fairs without RAM's support.

Another contributing factor was Virginia's expansion of Medicaid beginning in 2019. At the time of the final RAM clinic in Wise in 2019, more than 290,000 Virginians had enrolled in the program. Nearly a year later, that figure grew to more than 425,000, including nearly 4,000 people in Wise County — a little more than 10% of its total population — according to data from the Virginia Department of Medical Assistance Services.

As coal continues to decline as an economic driver, communities in central Appalachia and Wyoming are trying to determine their next steps in maintaining critical healthcare infrastructure.

But healthcare also remains a politically charged issue. The Patient Protection and Affordable Care Act of 2010, also known as Obamacare, has been a flashpoint in partisan politics since it passed. The law brought sweeping changes, including incentives for states to expand their Medicaid programs to cover citizens and legal residents with income up to 138% of the poverty rate.

Still, shrinking tax revenues, jobs and resources in the industry are forcing the issue in many parts of coal country, and communities are becoming more reliant on — and even warming to — federal health programs like Medicaid.

TO EXPAND OR NOT?

Only 14 states had not expanded Medicaid coverage by 2020, including Wyoming, where continuing layoffs and declining revenues from coal and oil threaten a rural health system that the U.S. News and World Report ranked 42nd in the nation for overall quality of healthcare.

Most states in Appalachian coal country have expanded Medicaid. Kentucky and West Virginia expanded Medicaid in 2014, the first year they could do so. Both have Republican-majority state legislatures, as well as governors representing both parties. Even with GOP efforts to weaken Medicaid, as well as failed attempts to impose work requirements on recipients, the percentage of uninsured people in both Kentucky and West Virginia is significantly less than in Wyoming, which has yet to expand Medicaid, or neighboring Virginia, which didn't expand until 2019.

Despite assaults by state lawmakers, Medicaid expansion has remained popular in Kentucky and West Virginia, to the point that no one's talking about legislative repeal any more. In 2018, incumbent U.S. Sen. Joe Manchin, D-West Virginia, who had famously shot the cap-and-trade bill in a campaign spot eight years earlier, reprised the ad — only now he was shooting a lawsuit to repeal the Affordable Care Act. The message helped carry the Democrat to victory in what had been Donald Trump's second-best state just two years prior.

"One thing that's struck me is how Medicaid in the last five years or so has become much more popular than it ever was," said Adam Searing, a research professor at Georgetown University's Center for Children and Families. "People are seeing it more like Social Security or Medicare — not so much welfare but a safety net program."

As a countercyclical program, Medicaid is designed to kick in when the economy suffers. That's been happening in coal country since the Great Recession, and it spiked across the country with the pandemic.

"Ideally, people want a job, but if the jobs aren't there, you don't want people to also lose their health insurance too, especially these days," Searing said. "States that have expanded Medicaid have more options for people to retain good health coverage than states that haven't. What this really means for families is financial security."

One of the commonalities of coal communities in Wyoming and Appalachia is they tend to be located in rural areas where there aren't enough people with adequate insurance to help cover the costs of healthcare services.

With coal in a downward spiral, that means more people in coal country — including an aging population of miners — are relying on social programs like Medicaid and Medicare. Coal states that initially rejected Medicaid expansion now see it as a way to help stem some financial losses in healthcare and provide care to residents. Increasingly, the argument that expansion is vital to aiding coal communities' transition to a sustainable post-coal economy is gaining traction.

"I think [Medicaid expansion] is actually a really strong conservative position," said Jen Simon, senior policy advisor for the nonpartisan Equality State Policy Center, which represents social justice, conservation and labor organizations in Wyoming. "There's great evidence across the country that Medicaid expansion actually results in more productive employees, better job creation and great economic growth. I would suggest that's all in keeping with Wyoming values."

A COAL STATE RELUCTANTLY RECONSIDERS MEDICAID EXPANSION

After repeatedly rejecting expansion efforts as recently as February 2020, Wyoming legislative leadership called on members the following May to again take up the question of Medicaid expansion in light of the coronavirus pandemic and historic losses in coal.

But just weeks after the call to add Medicaid expansion to the Wyoming Legislature's list of special session priorities, a committee failed to take action. Similar efforts to expand Medicaid in Kansas and Oklahoma also stumbled.

"I've been very surprised at the continued resistance to Medicaid expansion," Searing said. "I thought surely this would be the catalyst. But it's still an uphill battle to overcome this ideological resistance."

Wyoming's refusal to expand Medicaid coverage under the Affordable Care Act has cost the state nearly $1 billion in federal funds since 2013, as well as an opportunity to cover up to 32,000 residents by 2022 — a number that continues to grow with mounting job losses.

So far, it's been a price that the conservative-dominated Legislature is willing to pay in order to align with a staunchly anti-federal government voting base.

However, expansion proponents still hold out hope that lawmakers — particularly those from Wyoming's coal communities — might change minds in the Legislature. They say Medicaid expansion would soften the blow to both patients and medical providers in the short term, and it may be vital to sustaining hospitals and Wyoming's workforce in economic transition.

Legislative leadership in May 2020 cited recent job losses and ongoing uncompensated care when it voted in favor of studying expansion in response to COVID-19 and a state budget crisis. Wyoming hospitals lose an estimated $100 million annually in uncompensated care — a figure that was likely to spike amid the pandemic.

Rep. Tom Crank, R-Diamondville, represents a district in southwest Wyoming that includes several coal communities. He said he's mostly agnostic toward Medicaid expansion, but analysis from the Wyoming Department of Health seems to justify at least considering expansion. "For every dollar you put in, you get 10 dollars back. If that's the case, sign me up," Crank said.

However, voters in his district — as well as party leadership — have "set me straight" when he's brought it up, Crank added. "We're all kind of fat and happy as long as things are going all right," Crank said. "But if you start seeing one or two of these rural hospitals go down, that will be the red flag. They're really important to these small communities. We need them to make these communities viable."

COAL MINERS HAD IT GOOD, FOR A WHILE

When it comes to the rising cost of health insurance, coal companies are no different than any other big employer. As costs increased over the past two decades, so did premiums and deductibles for coal miners and their families. It's a trend that state leaders struggle to find politically acceptable strategies to counteract, and one that confounds a segment of high-wage earners who previously enjoyed robust health benefits.

Don Dorn worked at Powder River Basin coal mines for more than 30 years before taking a job with the state as a deputy mine inspector. When he hired on at Black Thunder mine in the 1980s, employees enjoyed private insurance with no premiums or deductibles. Black Thunder even had a full-time nurse practitioner onsite to perform physicals, prescribe medications and see employees who were ill — at no cost.

"It was like having a personal nurse right there at your fingertips," Dorn said. "It was an awesome benefit. It saved the company probably hundreds of thousands of dollars each year."

Eventually, premiums began taking small bites out of paychecks. Then bigger bites. After Arch Coal took over the mine from ARCO in the late 1990s, it eventually cut the nurse position. Things began to change dramatically in the 2000s. First there was a $100 deductible, then $500, then $1,000, along with in-network limitations. Now some miners in the Powder River Basin have deductibles in the $5,000 range.

The rising cost of healthcare hit coal companies and miners much the same throughout the state.

"I don't think the trend has been different from any other industry," Dorn added. "As insurance got more expensive, you know, they started putting more burden on the employees."

Dorn also serves on the Wyoming Miners' Hospital board. The hospital provides some basic services — mostly related to respiratory and hearing — as well as financial assistance to miners with more than 10 years of mining experience. There's been a steep rise in demand for services in recent years — from retired coal miners, those who have been laid off and even among miners still on the job, Dorn said.

"Most people never used it that much, and within the last five years we've almost reached our capped amount [of benefits] each year," he said.

Today, health coverage for Wyoming coal miners varies from "good" private coverage to not so good. Hundreds of retired coal miners in Wyoming and Appalachia temporarily lost healthcare coverage when Westmoreland Coal filed for bankruptcy in 2019. Later the same year, Blackjewel canceled healthcare coverage for thousands of employees in Wyoming and Appalachia as it spun into bankruptcy and reneged on obligations to pay salaries, retirement and taxes owed in coal country.

Healthcare was restored for thousands of union miners in late 2019 with the passage of the bipartisan American Miners Act, essentially a band-aid to the faltering health care and pension funds program established between the federal government and union miners. Wyoming's congressional delegation did not support the deal; union miners make up a tiny percentage of Wyoming's coal workforce.

LESSONS FROM MEDICAID EXPANSION IN APPALACHIA

For years, Virginia also resisted Medicaid expansion, with a Republican-majority Legislature that blocked the Democratic governor's efforts to expand. But in 2018, following an election in which Democrats came within a single, disputed vote of parity in the House of Delegates, a bloc of Republicans, mostly from the mountainous western part of the state, broke from their party and teamed up with Democrats to finally expand Medicaid.

Del. Terry Kilgore, an influential delegate from Virginia's southwestern corner who's served in the House since 1994, was the first key Republican to publicly call for Medicaid expansion. In an op-ed published in the Roanoke Times, Kilgore argued that Medicaid expansion was essential for the region's workforce.

In an interview that year, Kilgore said that coal's decline in his district played a major role in his decision to change his long-held stance against Medicaid expansion.

"When coal started its descension, it was No. 1, the job creator," Kilgore said. "But when you lost all the coal jobs, a lot of people lost their healthcare. People were working but were going to jobs paying $8 to $15 per hour with no healthcare benefits. Couple that with the hospital closure in Lee County. All that added together made me realize why we need to address this problem."

Nineteen of 51 Republicans in the House and four of 21 Republicans in the Senate broke ranks to join Democrats in expanding Medicaid. Republicans in coal-producing parts of southwestern Virginia played a fundamental role in the movement toward expansion, but the delegation split on the final vote.

The federal funding that came with expansion has not only been used to cover more Virginians, but also has freed up state money to pay for expanded mental health and other healthcare programs.

Meanwhile, free clinics and community health centers that helped fill the healthcare gap in Appalachian Virginia before Medicaid expansion are evolving to cover a population that is now better insured, but still in need of access to healthcare.

For the Health Wagon, Remote Area Medicine's longtime partner in the summer clinic at the Wise County Fairgrounds, innovating is a tradition. The Health Wagon was founded in 1980 by a missionary who operated out of a Volkswagen Beetle; today it operates at 13 sites in six counties and the city of Norton. Its clinic in Coeburn operates on a fee-for-service model — a break from the Health Wagon's longtime operation at a free clinic, but one which allows it to treat people with Medicaid.

The declaration of a pandemic in early March 2020 interrupted a series of day clinics in three communities, as well as regular operations, said Health Wagon Executive Director Teresa Tyson. Many of the Health Wagon's regular patients rely on it for ongoing treatment of chronic, long-term conditions such as diabetes, hypertension and COPD.

"People die without care," Tyson said. "It's important for providers and nurses to stay healthy, so we can continue to provide care, and on the flip side keep the patients safe and at home."

That's meant using telehealth connections where possible. The Health Wagon also keeps iPads in the lobbies at its three permanent locations, which allows patients without a phone to visit and connect with a nurse or physician virtually, even if they don't have the technology to do so at home.

Proponents in Wyoming say expanding Medicaid is necessary to staunch uncompensated care and free up funds needed to invest in telehealth. This will facilitate better coordination of care among a statewide system of varying capabilities to deliver services to where people live and work.

"That's how we improve healthcare in rural settings and get to reach those vulnerable populations, getting access to them," said Dr. Tracey Haas, a family physician clinical instructor at the University of Wyoming. "You don't do it just by setting up a clinic. You do it by actually going out and being with them and seeing what their actual needs are and meeting those needs where they're at.

"My students want to do this kind of work," Haas continued. "They want to get that in their training. But this is something that's hard to bootstrap and fundraise for. It's something that I think having an expansion program would help us do."

The pandemic eventually may result in systemic changes to America's healthcare system. In the meantime, on-the-ground hospital systems and clinics such as the Health Wagon continue to make do through innovation and experimentation.

"We're meeting our patients where they're at," Tyson said. "We're resourceful. We're creative people. One of the biggest problems with southwestern Virginia has been that singular, extractive economy. If there's a lesson to be learned, it's don't put all your eggs in one basket."

Photo by Dustin Bleizeffer / WyoFile

Gillette, Wyoming, has long served as the hub of the Powder River Basin coal complex, which supplies about 40% of the nation's thermal coal for power generation. Mine workers and local businesses have scrambled to adjust to a coal industry downturn that may only get worse.

CHAPTER FOUR

RECKONING IN COAL COUNTRY

Lax fiscal policy has left states flat-footed as mining declines. What happens when a $28.6 billion industry spirals into permanent decline?

Rose Evenson's mining roots run deep.

Her grandfather and uncles worked at the Homestake gold mine in Lead, South Dakota. As Homestake wound down, a lot of its miners went to work at Powder River Basin coal mines just across the border in Wyoming.

In the 2000s, Evenson and her brother followed and took jobs at Black Thunder mine — about a 1.5-hour commute from her South Dakota home in the Black Hills.

"I thought I'd be a coal miner until I retired, but things changed," said Evenson, 55. "It got stressful because you were always wondering, 'Is this the month? How long do I have here?' Every year it kept getting slower and slower, and I kept thinking, 'God, I got to do something else.'"

After 13 years at Black Thunder, Evenson took a buyout package in spring 2020. Many of her colleagues did the same and moved to places like Utah, Indiana and Missouri. Evenson worked part time that summer for the local parks department in South Dakota while training for her commercial driver's license.

"[Black Thunder] was a great place to work," Evenson said. But job prospects elsewhere in the region are bleak. "There ain't nowhere in this whole area — South Dakota, Montana or Colorado — [where] if you ain't working for a coal mine you ain't making much money. But I don't see coal mines booming anytime soon."

Evenson's assessment of the industry's future stands in contrast to legislative action in coal states, where for decades a dogmatic belief in coal's longevity has guided fiscal policy that has in some cases actually made it harder for coal-region economies to diversify.

And while thousands of mining jobs are being lost around the country, coal's collapse carries ramifications that reach far beyond coal towns themselves, affecting downstream industries with larger geographic footprints. Railroads, for example, are slashing jobs along coal routes in response to declining shipments between coal mines and the power plants they serve. Manufacturers of equipment used in the coal industry have taken a hit as well.

So what happens to communities in coal-producing regions when a $28.6 billion industry spirals into permanent decline?

High-salary workers like Evenson either retire earlier than planned or search for another, most likely lower-wage, job. Some move away and many become more reliant on social health services.

Businesses lose customers and healthcare providers see fewer patients with adequate insurance. Charitable giving among businesses to support local nonprofit social services dries up just as the need for such services skyrockets. Locally and regionally, revenues to support government services plummet, triggering budget cuts — often to the very programs most needed to maintain a quality of life and transition to more sustainable economies.

Wyoming also disproportionately relies on tax revenue from mineral extractive industries to fund education. Since 2016, the state has made cuts and dipped into savings to help staunch a growing K-12 education budget shortfall. The situation only looks to intensify.

Part of Wyoming's emergency budget-cutting in the energy downturn and COVID-19 pandemic — the state slashed $250 million by summer 2020 while budget officials eyed tripling that amount in the coming year — included an undetermined number of layoffs among teachers and state workers, according to Gov. Mark Gordon. The job losses would hit local economies everywhere in the state, Gordon said. The revenue picture in Wyoming was so bad in the summer of 2020 that if all of the state's roughly 7,500 state employees were laid off immediately, the savings still would not cover the estimated $1.5 billion biennium budget deficit.

"We spend every capital gain we get. We are living from hand to mouth as a state," Gordon said in his opening remarks during a 2020 public forum on the state's tax and revenue picture.

In many ways, 2020 was the reckoning that Wyoming and other coal communities never thought possible. Analysts predicted closures among coal mines once considered "crown jewels," and the state's revenue was in freefall.

"When I started, we honestly believed coal was a 200-year asset," said Sen. Cale Case, a Republican from Lander and an economist and 25-year veteran of the Wyoming Legislature. "Nobody has planned very well for what happens post-coal."

A TAX STRUCTURE BECOMES A LIABILITY

Over the past 50 years, Wyoming has built a tax-and-revenue structure that not only narrowly relies on extractive industries to fund basic government services, but actually works against any economic diversification beyond extractive industries.

Wyoming's structural tax-and-revenue conundrum can be summed up in this dataset: The average household of three with an income of $60,000 and a home valued at $200,000 in Casper paid $3,070 in state taxes in 2017 but received $27,500 in public services, according to state data compiled by the Wyoming Taxpayers Association. That is the result of the state's reliance on mineral extraction to pay the bulk of government services — about 70% of the state budget in 2000 and 52.2% of the state budget in 2017.

So even if Wyoming is successful in growing its non-mining industries, those new businesses and employees will be more of a drain on the budget than a contributor to it without reforms to the state's tax and revenue structure.

"This is just a difficult question because they're not going to replace coal [revenue]," Wyoming Taxpayers Association Executive Director Ashley Harpstreith said. "Until we change our tax structure, there's no way to replace that revenue."

This isn't the first time a commodity downturn has forced Wyoming leaders to consider reforming the state's tax-and-revenue model. For example, the Tax Reform 2000 report was spawned by a prolonged downturn in energy in the 1990s. But the report's key recommendations, which included imposing corporate and income taxes and closing a series of exemptions, were quickly shelved when the next boom came along: coal-bed methane gas in the early 2000s.

For decades, lawmakers have sidelined discussions about recalibrating lodging and gasoline taxes, as well as considering a corporate or personal income tax. However, there's renewed interest in reexamining Wyoming's Tax Reform 2000 report as a starting point for potential reform. The Wyoming Taxpayers' Association hosted an online forum, "A Twenty Year Review of Wyoming Tax Reform 2000," in July 2020 that featured several experts and elected officials.

But whether Wyoming officials — and the voters who elect them — have the appetite to impose new taxes on themselves remains a big question.

Sen. Eli Bebout, a Republican from Riverton, said he believes, even in light of the COVID-19 economic crisis, Wyoming has more of a spending problem than a revenue problem, alluding to a longstanding strategy among lawmakers to "cut-and-cope" during energy downturns.

Former Sen. John Hines, a Republican from Gillette who led the Tax Reform 2000 effort, said he sees little evidence that Wyoming is prepared to take on tax reform. "I've been asked a lot in the last 20 years, 'When is Wyoming going to do something different?' I say, 'When all of our savings is gone.'"

COAL LOSSES HIT BASIC SERVICES

The ways in which communities receive revenue from the coal industry vary by state. Kentucky, Virginia, West Virginia and Wyoming have different approaches for collecting, sharing and distributing coal-related taxes to localities. However, all four use a state severance tax. Local governments in all four rely on property taxes as a primary revenue source. All of these revenues are declining.

Take Kentucky: The state coal severance tax historically has gone to local government through two funds, focused respectively on economic development and economic assistance.

"The theory was those resources were supposed to be used to create infrastructure so that when coal's depleted, those areas can continue to thrive," said Pam Thomas, senior fellow at the Kentucky Center for Economic Policy. "The reality is, those were used for everyday expenses, and things like sports stadiums, senior citizen centers and planting flowers in medians. They did do some industrial parks, but didn't have the roads and broadband and other infrastructure needed to support those parks. A lot of the time, those parks were created and sat empty."

Revenue sources including personal property and gas taxes have declined, too.

"We had a big structural deficit at the state level in Kentucky before COVID happened," Thomas said. "The rainy day fund at the state level is almost zero. The capacity of the state in general to weather the storm we're in now is diminished. The state doesn't have many resources to help local governments. Absent federal aid, there really isn't a lot of help."

Coal severance collections have plummeted. In Letcher County, Kentucky, for example, tax receipts fell 89% from 2009 to 2019, and the county's workforce fell by 25%. That decline squeezed the county's budget, which shrunk by 40% between 2012 and 2018. The county tried to increase revenue by taking more state prisoners into its local jail, with the result that the jail has become one of the most overcrowded in Kentucky, a problem that became even more consequential during the pandemic.

Elsewhere in Kentucky, local governments have enacted partial government shutdowns or cut into basic services such as waste management and law enforcement. Many places have failed to maintain and build infrastructure that's necessary for an economic transition from coal, resulting in failing water systems, flooding and other crises.

"All of the water in the coalfield areas is impaired by mining," said Mary Cromer, deputy director of the Appalachian Citizens Law Center. "It's so pervasive, it's hard to say exactly how it affects growth or development or the chance for anything else around here. It puts a strain on water treatment systems."

Water isn't the only problem. As broadband internet became even more important for working during the pandemic, Kentucky lagged the rest of the country in high-speed internet access, in part because a $1.5 billion initiative to wire the state was running behind schedule and over budget. Infrastructure challenges make economic development more difficult in a region where rugged terrain and relative inaccessibility pose inherent obstacles.

MINE RECLAMATION: AN OPPORTUNITY AND POTENTIAL RISK

Appalachian communities are struggling not just with the effects of coal's decline on revenues, but also on the land. A 2016 Duke University study showed that extensive mountaintop mining, using techniques that blow the tops off ridges and deposit the debris in nearby valleys, has left parts of central Appalachia 40% flatter than before. Toxins from those mine sites are leaching into waterways, Duke researchers found.

Mine reclamation, required under the Surface Mining Control and Reclamation Act of 1977, has the potential to replace some lost coal jobs while preparing the land for new uses. But it also represents a massive liability with a history of misuse. More recently, reclamation has become a liability that the coal industry is deftly shirking through bankruptcy and the assistance of policymakers desperate to help mitigate the industry's financial woes.

Between 2012 and 2017, Alpha, Arch Coal, Patriot Coal and Peabody Energy shed nearly $5.2 billion owed for reclamation and miner benefits, in part by spinning them off to other companies, according to a Stanford Law Review study.

Blackjewel, which held mining permits in Kentucky, Tennessee, Virginia, West Virginia and Wyoming, acquired more than 80% of its holdings from previous coal bankruptcies, including by Alpha Natural Resources and Arch Coal. In 2019, Blackjewel itself filed for bankruptcy, which played out in chaotic fashion and saw the company trying to jettison its own, largely secondhand, obligations. Its more productive mines and facilities were picked up by other companies, who simultaneously jockeyed to avoid its unreclaimed, unproductive mines. All this time, those unreclaimed mines have racked up reclamation violations.

"What we're seeing is the inevitable end game of the industry here," said Willie Dodson, central Appalachian field coordinator for Appalachian Voices. "What we're seeing with the Blackjewel bankruptcy is this large number of permits that no reasonable person would expect to make money out of, different entities trying to shove them off on one another, and state regulators and insurance companies increasingly satisfied to just have any coal company on the hook, just to shield themselves from having the reclamation obligation."

Wyoming strengthened its bonding and surety requirements for mine reclamation in 2019 in response to coal company bankruptcies and moves to shed their cleanup liabilities. But taxpayers in the state could still get stuck with huge cleanup costs if a handful of companies forfeit bonds and walk away from their reclamation obligations.

Stacy Page, board member of the landowner advocacy group Powder River Basin Resource Council, said she worries that as profits shrink, coal companies will fall further behind on concurrent reclamation work, or lay off workers on reclamation crews.

State regulators have been lenient in enforcing reclamation, looking for ways to let coal companies catch up on payments for violations instead of forfeiting the bonds set aside for reclamation should the company default, according to advocacy groups. The alternative is letting those costs fall on state taxpayers.

That's because, critics say, most state systems for setting bonds are structurally flawed. In some cases, states allowed companies nearly $4 billion in "self bonds," or guarantees that they had the money to reclaim mines. The majority of that amount was for companies that eventually went bankrupt. Many states no longer allow self-bonding.

Kentucky, Virginia and West Virginia still all use bond pools, a collective fund paid into by coal operators. Bond pools work like insurance; if a mine operator goes under and can't pay for reclamation, the difference is paid from the bond pool.

However, all three central Appalachian states face situations where one operator's bankruptcy threatens to completely destabilize their bond pools. In Kentucky, it's Blackjewel. In Virginia, it's companies that belong to the family of West Virginia Gov. Jim Justice. In West Virginia, it's mines owned by Tom Clarke, who feuded with Justice before becoming a coal baron himself.

Their mines sit unreclaimed, scarring the landscape and leaking pollutants. They act as anchors slowing attempts by coal communities to build new economies.

"If mines just sit unreclaimed because nobody is in a financial position to do that, the risk of flash floods and landslides and debris being washed off-site is just going to become worse," Dodson said. "Any long-term treatment that is needed for water quality is not going to be happening. It's pretty hard to keep your young people if places are dangerous and polluted. It's certainly difficult to attract investment in that kind of place."

Although surface coal mines in the Powder River Basin generally have a better environmental track record than historic and mountaintop removal mining in Appalachia, they still pose significant environmental and health risks if neglected.

Groundwater is a critical resource on the high arid plains, and in the past, mines have exceeded dust standards — a respiratory hazard. A surface mine in Wyoming, if left abandoned, could fill with toxic water, and leave hazardous highwalls and coal seams that could catch fire and burn underground for decades. Properly reclaiming disturbed surface in the Powder River Basin is essential to ongoing agricultural operations and the region's wildlife.

According to a 2020 report from the Western Organization of Research Councils, the reclamation obligation for the thousands of square miles of surface mines in the Powder River Basin adds up to nearly $2 billion — which represents a financial and jobs opportunity that can help miners and service companies just as mines are closing. Coal Mine Cleanup Works, another 2020 report by the Western Organization of Resource Councils, suggests reclamation at surface mines in five western states could create between 4,893 and 9,786 full-time jobs, mostly in Wyoming.

No coal community can afford to squander those reclamation opportunities, said Kate French, regional field organizer for the Western Organization of Resource Councils. She said there's a tendency among policymakers to ease standards on the industry in the hopes and promise it will extend how long mines can operate, but it comes at a huge risk.

"That approach to helping the industry and, ostensibly, to help out the workers is rooted in something that is no longer the reality today," French said. "It doesn't matter how many regulations you cut or how many incentives you give the companies — the market doesn't want [coal]. It's not going to change what's happening in energy in the nation and around the globe."

"I think a lot of coal companies like to obscure where that [reclamation] money comes from," French continued, adding that it comes from coal companies — a cornerstone to the deal they struck with the American people to mine coal in the first place. "They've been making money hand over fist for quite a while; why not give back to the communities that have really supported these coal companies and invest in them and invest in job creation?"

Rafters at New River Gorge.

CHAPTER FIVE

PATHS FORWARD

Communities contemplate how to survive a post-extraction economy in manufacturing, reclamation and renewables.

The blue rubber rafts shriek as they slide down steep metal rails, each guided by a crew that will soon be floating down a 9-mile stretch of river, replete with Class III, IV and V rapids and the ghosts of now-shuttered coal mines and processing plants.

Raft after raft descends the skids from a parking lot for commercial outfitters before arriving at the put-in, where guides instruct their crews of clients. The churn of activity is a little lower than what it might typically be on a hot day in late July, but it's still surprisingly busy for a Monday morning in the midst of the novel coronavirus pandemic.

The steady flow of boats into West Virginia's New River also exemplifies the progress that the outdoor adventure industry has made toward diversifying the regional economy around the river.

Signs of the industry's past in the New River Gorge can still be seen at Kay Moor, where remnants of a coal tipple, coking furnaces and an iron frame reading "Your family wants you to work safely" can still be seen in the forest above the river.

Fayette County produced 3.2 million tons of coal in 2019, about 3% of West Virginia's total, but today the county is known less for coal than its plunge into outdoor adventure, beginning with the rafting industry that emerged in the '60s and '70s as coal was winding down. Outfitters and adventurists were attracted by access to the gorge and surpluses of cheap housing and store-fronts in the wake of coal's post-World War II decline.

The New River Gorge became a magnet for rafters, climbers, hikers and mountain bikers, to the point that West Virginia historian John Alexander Williams included it as an example of a possible economic path forward in his comprehensive 2002 history of Appalachia, citing "the growth of whitewater rafting and other recreational activities that have changed the New River Gorge from a transportation barrier into a tourist attraction."

Tourism creates seasonal jobs and economic activity, not just among outfitters and guide services but in restaurant, lodging and other service businesses.

But "that in and of itself isn't enough to create resilience in the community," said Kelly Jo Drey, a part-time raft guide and former Fayette County employee who worked on turning recreation into economic development. "The thing that outdoor recreation provides for the community that's more far-reaching in its impact is the ability of those amenities to bring people here and keep people here for other reasons."

As the COVID-19 pandemic pushes more people to work remotely, communities like Fayetteville hope to attract new residents who can do their jobs while living in a beautiful region in close proximity to recreation opportunities. In doing so, the town is competing against numerous other coal communities trying to emulate its relative success by using outdoor recreation as one strategy to diversify their economies.

Dozens of communities in the Appalachian coal economy have been plotting a course beyond coal for decades, relying on local culture, outdoor recreation, tech jobs and a skilled labor force to expand their economies.

In Wyoming, where coal's downturn has been more recent and abrupt, initial responses to expand and diversify have played to "readily acceptable" jobs and businesses that match the state's culture and labor force, including coal technology research and recruiting gun manufacturers.

In a move driven by state-level policies, gun manufacturer Weatherby moved from California to the mining town of Sheridan, a community at the foothills of the Bighorn Mountains expanding its reputation as a gateway to Yellowstone and the West. Other firearms manufacturers have also found a warm reception in Wyoming: Magpul and Stag Arms both relocated operations to Cheyenne — citing a like-minded gun culture and policies.

Tourism and outdoor recreation play a prominent role in Wyoming's economy — No. 2 to mineral extraction in terms of revenue to the state while dwarfing minerals in terms of jobs. More than half of the state is public lands, including six national parks, eight national forests and 13 state parks. Its population density is six people per square mile, second-lowest behind Alaska. The COVID-19 pandemic made these factors a draw to Wyoming beyond tourism and outdoor recreation, according to economic development officials.

"I really do believe the work that people do remotely [during the pandemic] will accelerate plans for people to do more remote work," said Wyoming Business Council Executive Director Josh Dorrell, adding that the council is actively working to recruit more tech and information-based businesses to work remotely in the state.

But the Powder River Basin's wind-swept grasslands are a long way from Yellowstone and the Tetons, both figuratively and literally. And despite these efforts, there's no interchangeable roadmap to the essential challenge of rebuilding rural economies once dominated by coal.

The damage done by decades of reliance on the coal industry — unreclaimed mineland, depressed local revenues, depopulation and environmental degradation — makes it difficult to attract new economic development in Appalachian regions, especially to places that aren't located along major transportation corridors. Yet for many, moving isn't a viable option.

"People care about their place," said Tom Hansell, a documentary filmmaker, author and artist who teaches at Appalachian State University and made the film and book "After Coal." "In some cases, people don't have a lot of options to move — their educational background or financial freedom didn't allow them to move and find employment. But there's also an intense connection to place and culture that's tied into identity. That's what kept people in those places and provided a foundation — not a financial but a community foundation — to survive."

PLAYING TO ONE'S STRENGTHS

Keith Kinsinger spied through the observation glass of a whirring Deckel Maho DMU 200 FD — a five-axis automated machining "robot" and one of the most sophisticated pieces of equipment at L&H Industrial's expansive welding, machining and manufacturing campus in Gillette. He was using the Deckel Maho to cut metal teeth on a large industrial gear.

"I've had my eye on this machine ever since I started here," Kinsinger, 21, said. "If you can dream it, this machine can do it."

Kinsinger worked a stint in the oilfields after high school but worried about job security. A friend recommended machining courses.

"I thought that was a good career choice and I started going to college," he said. He enrolled at Gillette Community College, then finished up at Northern Wyoming Community College in nearby Sheridan. L&H regularly recruits students from local trades programs to find talent like Kinsinger.

The Gillette native said he feels secure in his job, even considering the energy downturn.

"We're still keeping a steady 40 hours a week," Kinsinger said. "I think it's a good career. Hasn't let me down yet."

The declining Powder River Basin coal industry — once a lifeline for L&H — no longer threatens the company's viability or ability to keep growing in the heart of Wyoming coal country.

"Even if there wasn't anything to do here in Wyoming, I believe that the 200 [L&H employees] that are here would stay here working and getting paid the same, and everything would just be shipped to somewhere else in the world," L&H Industrial President Mike Wandler said.

If the key to diversifying against the cyclical boom-and-bust nature of an extraction economy — or even the loss of a major economic driver — is to play to one's strengths, then L&H is a prime example in Wyoming.

In 2013, coal mining clients made up 30% of L&H's revenue. Now it's 10%. Yet the company's financial position is stronger than ever, mostly because 60% of its customers are outside Wyoming and across the world. L&H started out in 1964 as a small, family-owned welding shop serving the oil and gas industry in northeast Wyoming. It expanded to serve the region's coal mines as a way to sustain itself during oil and gas downturns. But even that level of diversification doesn't ensure sustained success, Wandler said.

It became obvious that L&H's innovations to save money for coal mines in Wyoming could be exported to mines across the world, Wandler said. The next step was to build on its expertise in advanced manufacturing.

"In Gillette, Wyoming, you can create a heavy industrial business that sells mostly out of Wyoming and out of the country, and that's where you get your diversity," Wandler said. "They don't need to completely reinvent themselves, but you have to be selling into the world economy and you have to be figuring out how you can be relevant in the world."

Wandler sees untapped potential for manufacturing in Wyoming. He served on the Wyoming Business Council board of directors, and also played key roles in former Gov. Matt Mead's efforts to drive economic diversification in Wyoming. Many coal, oil and gas service companies can branch into manufacturing, including aerospace and defense contracts, Wandler said.

Local leaders in Sheridan County, which also serves the Powder River Basin coal industry, redoubled efforts to diversify after the coal-bed methane gas industry went bust in 2010. Since then, Sheridan County has expanded jobs in tech and manufacturing, including Kennon Products, which manufactures sun shields and other products for government and private aviation. Other local service companies have expanded beyond oil, natural gas and coal to tailor to wind energy customers; wind developers are poised to invest some $10 billion in Wyoming.

ADVENTURES IN REINVENTING MINELAND

In parts of Appalachia, coal has been joined, though not replaced, by a burgeoning industry that seeks funding for schemes to repurpose land on and around former mines into new economic endeavors. These ideas range wildly, but often involve outdoor recreation, heritage tourism, agriculture or renewable energy.

Often, these proposals seek nonprofit grants or federal funding from the more than $300 million that Congress has appropriated for the Abandoned Mine Lands Fund that was established through the Surface Mining Control and Reclamation Action of 1977. The nonprofit advocacy group Appalachian Voices produced reports in 2016 and 2018 in collaboration with a number of other organizations that promoted redevelopment of dozens of sites throughout central Appalachia, including several that eventually won Abandoned Mine Lands funding.

The federal funding has also been used to support the development of federal and state prisons, which have become major employers in several coal communities. Not all of those prison projects have found success.

Other projects aiming to reinvent former mineland have fallen as well.

In 2016, a plan to farm lavender at a reclaimed strip mine in Boone County, West Virginia, won grants from the Benedum Foundation and the Appalachian Regional Commission, then abruptly collapsed when the grants ran out. Similarly, the non-profit Mined Mines promised to train and place people in coding jobs, but failed to follow through on its promises, leading to accusations of fraud.

WHAT'S WORKING

Appalachia has seen successes, often small and built over time. Fayetteville, West Virginia, has the advantage of its location near the New River Gorge National River, a unit of the National Park Service, while other communities market attractions in national forests and state parks. Both Kentucky and Virginia have built networks of off-road trails modeled after West Virginia's Hatfield-McCoy Trails, which wind more than 700 miles across 14 counties in the state's southern coalfields.

Many communities are building hospitality industries around outdoor adventure in hopes of attracting not just visitors, but talented workers and companies that want to hire them. Making that happen requires infrastructure — and not just basics like good water and reliable electricity, but broadband internet, healthcare access and higher education to train an ever-evolving workforce. In places struggling to maintain basic government services, that's a tall order.

Even the effort to find solutions has proven beneficial.

"One of the things that really gives me hope is that people are more willing than ever to come together right now — despite what their politics are, despite what they believe in — and think about how they rebuild a thriving community that considers all of the people who live there, all of the people who may visit there, and all of the aspects that are unique to that particular place," said Ivy Brashear, Appalachian transition director for the

Mountain Association. "Previous to the past decade or so, we hadn't seen that in a way that's been able to catch fire and spread across the region. It's really powerful."

Those conversations have created a new energy among groups like the Letcher County Culture Hub and What's Next EKY?!, which seek to empower locals to advance their own ideas. Combined with the ability to connect on social media, people are finding mutual support that is slowly changing the internal narrative about coal country.

"What we're seeing with these groups is that young people are understanding they can live the kind of life they want to live while staying in place," Brashear said. "A lot of young people are staying in the region and investing in their own communities."

CAN COAL TECHNOLOGY AND RENEWABLE ENERGY HELP FILL IN THE GAPS?

Wyoming taxpayers have invested tens of millions of dollars over the past 15 years to help advance technologies to reduce coal's greenhouse gas emissions — a significant driver of climate change.

The potential for advancements in capturing carbon from coal is frequently mischaracterized and misunderstood as a panacea for Wyoming's coal industry. It is not. If the goal is to save or extend the life of the Powder River Basin coal mining industry, the window of opportunity is quickly slamming shut.

That's because the future of Powder River Basin coal is tied to the U.S. coal-fired power plant fleet, and that fleet is quickly being retired, experts agree.

"We recognize that the clock is ticking," said Jason Begger, executive director of the Wyoming Infrastructure Authority, which oversees the Wyoming Integrated Test Center — a coal technology incubator launched in 2014 to "scale up" coal technologies for commercial deployment. "If we don't make some pretty giant leaps in scaling up and commercializing these technologies over the next five to 10 years, I think the [U.S. thermal coal] market is sort of going to pass us by."

Not all coal technology and research accomplishes the same goal. Begger said some of the Integrated Test Center's research is aimed at post-combustion strategies that might be applied at some U.S. coal-fired power plants — the most urgent need for Wyoming's coal mining industry. Other research, however, aims at pre-combustion strategies for yet-to-be-built coal plants — most likely outside the U.S. — or for smaller boutique markets for coal-derived products such as building materials.

The latter has become a major focal point for the University of Wyoming School of Energy Resources, which now has an annual budget of about $10 million.

Wyoming may benefit from the eventual commercial deployment of coal-conversion technologies for building materials at home, or even pre-combustion applications overseas, Begger said. But time is quickly running out for a post-combustion technology fix to help preserve today's U.S. power market for Wyoming coal.

"At some point there will be a tipping point where [U.S.] utilities feel like reinvesting in their coal fleet probably isn't a direction they're going to head, and that's just kind of the cold hard truth," Begger said.

Central Appalachia also is looking for ways to prolong the fossil fuel industry. Virginia lawmakers approved legislation to kickstart energy research in the coalfields for ideas such as emissions control strategies and using mine pools to cool data centers.

Appalachia's elected officials also are bullish on a plan to build out a new petrochemical industry in the Ohio Valley of western Pennsylvania, West Virginia, Ohio and Kentucky. The idea is to extend the shale gas boom of the early 2000s through the manufacture of natural gas byproducts such as ethane, which is used to make plastics and chemicals. A 2020 report from the U.S. Department of Energy trumpeted potential for growth "at a scale not seen since the Industrial Revolution" through "energy resource production, next generation manufacturing, and petrochemical industry development and expansion."

Yet that window may be closing as well. The same month the report was released, energy analysts warned that natural gas fracking firms face an impending wave of bankruptcy reminiscent of that which has devastated coal in recent years. For Appalachians, it's an all-too-familiar pattern.

Wind energy could potentially play a significant role in Wyoming's transition away from coal. Since 2014, wind energy developers have proposed investing nearly $10 billion in the state, which ranks among the top 10 in the nation for onshore wind energy potential.

For context, cities across America clamored at the opportunity to host Amazon's second headquarters, a $5 billion windfall. Power Company of Wyoming's 3,000-megawatt Chokecherry and Sierra Madre Wind Energy Project in south-central Wyoming is a $5 billion project. PacifiCorp, which operates as Rocky Mountain Power in Wyoming, plans to spend nearly $4 billion in Wyoming on wind, transmission and battery storage in the state.

"There is no other sector now, anywhere, that is thinking about a $4 billion investment in Wyoming," University of Wyoming energy economist Rob Godby said. "This is actually a great chance to diversify the economy and move to a fuel-secure outcome where we could be producing electricity for another several decades."

Wind, however, is a complicated topic in Wyoming. Despite the fact that ongoing wind energy buildout and upgrade projects in Carbon County alone have boosted local sales and use taxes there 215% during the past year, many lawmakers see any gains for wind as a net loss for Wyoming coal. The Wyoming Legislature continually reexamines how the state will tax the industry, including measures that some wind developers say threaten Wyoming's potential for wind development.

"We need certainty and stability in tax policy," said Kara Choquette, communications director for Power Company of Wyoming. "Our project has been under development for 12 years in Wyoming, and only four of those years did wind tax policy not change, or there wasn't the threat of a change."

Photo by Dustin Bleizeffer / WyoFile

Haul trucks at the Black Thunder mine in Wyoming prepare to dump a payload of coal into a "hopper," which will crush the large chunks and deliver the raw fuel to a train silo via conveyor.

CHAPTER SIX

SURVIVAL IS ANYTHING BUT CERTAIN

Coal country is not without options. But coal's long legacy of hope, promises and failure has instilled a political inertia that won't soon be overcome.

The 28-year-old man clearly had opinions on the future of coal, but wouldn't share them until out of earshot of a group of unemployed miners at a gas station in southern West Virginia.

"Economically, these coal fields are dying," he said, adding that he personally was done with coal. He'd finished three years of college before going into the mines, and was considering going back to become a teacher.

That conversation took place in 1956. It was recounted by Howard B. Lee in his book, "Bloodletting in Appalachia," a history of West Virginia's mine wars.

Central Appalachia has wrestled with decline ever since. It's hard to see the end of coal when it's been dying for 70 years.

The industry appeared to have a longer life ahead in Wyoming's Powder River Basin, but the last decade has demonstrated otherwise. Yet despite a steady flow of data and warnings from experts, obscurity still exists, delaying a sense of urgency to prepare for drastic changes.

In Wyoming, every bust has been followed by another boom in one fossil fuel or the other. Politicians promised voters that coal would drive wealth in Wyoming for generations to come. Now coal's demise is met with consternation and confusion. Even as it downsizes, the Powder River Basin will remain the cheapest-to-mine thermal coal region in the U.S. with the potential to pick up new supply contracts as higher-cost thermal coal mines close elsewhere. Wyoming is expected to be the last U.S. thermal coal supplier standing in a permanently shrinking industry.

"Seems like everybody is going on with their life," said Mike Wandler, president of Gillette-based L&H Industrial, the international manufacturer that also serves the coal, oil and natural gas industries. "There's still trucks flying off the lots, people are buying things, there's not a big fire sale. I guess I would say I've seen it a lot worse than this. I've seen it where you couldn't rent a U-Haul because everybody was loading them up and heading out. And that's not happening right now."

Both regions stand at a crossroads, faced with the decision to lay the groundwork for a post-coal economy or continue to lean into the fossil fuel-powered economic engine that propelled them for decades.

As conventional wisdom goes, hope is not a strategy. The coal industry is still contracting in both regions despite the fact that coal's political allies hold power on local and federal levels.

In West Virginia, resort and coal baron Gov. Jim Justice and state lawmakers cut the thermal coal severance tax. And still.

"None of the promises that coal was coming back happened," said Sean O'Leary, a senior policy analyst with the West Virginia Center on Budget and Policy. "There's no turnaround. This is a structural decline."

NO OFF-RAMPS

Appalachia produced the bulk of the nation's coal before it began to fall off in the 1970s; it was surpassed by the Powder River Basin around the turn of the millennium, according to data compiled by the U.S. Energy Information Administration. Along the way, central Appalachia faced a series of moments when it seemed it might diverge from its reliance on coal.

The passage of the Surface Mining Control and Reclamation Act of 1977 marked a potential reset point — when the cost of cleaning up abandoned mines would be covered, and new regulations would ensure the reclamation of surface mines in the future. In retrospect, the law essentially normalized the more environmentally intensive techniques of mountaintop removal mining and handed the responsibility of enforcement to state regulators, who just as often worked to help the industry as watchdog it.

In 2013, U.S. Rep. Hal Rogers, a Republican and longtime friend of coal, endorsed an initiative to develop a new economy in eastern Kentucky just a year after he had blasted then-President Barack Obama's so-called "war on coal." Eastern Kentucky's prospects have only grown bleaker as coal has continued to wane.

The 2019 miners' blockade of a train carrying $1.4 million in coal for bankrupt operator Blackjewel seemed to mark yet another moment of cultural shift. Some of the miners went back to work for other coal companies, and one operator that received assets in Blackjewel's bankruptcy has now filed for bankruptcy itself.

In Wyoming, Blackjewel's sudden bankruptcy resulted in a months-long pause in the operation of Eagle Butte and Belle Ayr mines, which threw miners' lives into chaos, but ultimately resulted in no permanent closure of major mining operations in Wyoming.

That lack of a clear rupture in Wyoming coal might contribute to the ongoing political inertia to plan for coal's structural decline.

University of Wyoming economist Jason Shogren said that with every challenge to coal, Wyoming chose to increase its wager on the industry's future.

"We knew this was coming 25-30 years ago," Shogren said. "We knew that there was going to be a push on reducing fossil fuels for climate change with the Kyoto Protocol back in 1997, and people in the state just decided to run with it. They decided to play defense, and now that day is here."

Wyoming Gov. Mark Gordon's policy director, Renny MacKay, told WyoFile in a 2019 interview that the state is focused on the need for economic diversification, and it is prepared to help communities struggling with the downturn in coal. However, MacKay said, the governor makes an important distinction when it comes to discussions about communities facing the challenge of an energy transition. "We don't really call it transition," MacKay said, "because the communities aren't looking to transition in Wyoming."

Wyoming isn't participating in policymaking that embraces a purposeful shift away from coal, MacKay said. It's fighting against those forces.

In the Interior West, Wyoming is the only state with no renewable portfolio standard or goals for reducing carbon emissions other than Idaho, which already gets most of its electricity from hydropower. In early 2020, Wyoming joined Montana in jointly petitioning the U.S. Supreme Court to overturn Washington state's regulatory roadblock to a proposed coal port expansion on the Columbia River — key to shipping more Powder River Basin coal to Asian markets. Gordon said he'll continue pitching Wyoming coal to potential Asian buyers, supported by $1 million in Wyoming taxpayer dollars.

Gordon signed into law a mandate forcing utilities to seek a third-party buyer for coal units in the state that they want to retire ahead of schedule. The state launched an investigation to challenge Rocky Mountain Power's economic and market analysis used to justify its plans to speed up the retirement of coal-fired power plants.

Perhaps the biggest factor when it comes to efforts to transition, for both Wyoming and Appalachia, is whether voters will continue to endorse efforts to save coal or help coal-dependent communities move beyond it.

TRANSITION TOOLS

Wyoming has several resources at hand to help itself in the decline of coal. More immediate mitigating tools include decommissioning and jobs programs among utilities that plan to retire coal-fired power units in the state, as well as coal mine reclamation — a potential $2 billion endeavor in the Powder River Basin. Wyoming also has more than $20 billion in savings and investments — some of which could be used to more directly help coal communities.

"If we're thinking about hard times and any kind of insurance package to help smooth out the income stream over half a century, a lot of attention should be paid to the sovereign wealth fund," Shogren said. "If people in Wyoming were more aware of what they actually have in [state savings and investments], we might be more aggressive in how we choose to invest those resources."

A 2020 analysis suggests the state can give itself more leeway in how it invests its sovereign wealth. "Everyone agrees that the downturn in the minerals industry is bad news for Wyoming," author Ben Gose wrote in a series published by WyoFile. "But Wyoming's weak investment returns also represent a significant — and rarely discussed — reason the state is going to have to cut spending or raise taxes."

States actively seeking coal transition strategies, such as Colorado, are looking toward securitization. It's a refinancing tool that can help reduce the ratepayer impact of retiring coal units early. Portions of savings from securitization go toward renewable energy and community development projects, which can in turn attract additional funds from the federal government.

Grassroots nonprofit groups such as the Powder River Basin Resource Council , Appalachian Voices and others have generated a font of ideas for assisting communities in transition from coal.

A range of local, tribal and labor leaders from coal communities across America endorsed the National Economic Transition (NET) Platform, developed through a process led by the Just Transition Fund. The platform outlines principles and processes, but largely leaves specific details to be developed by local communities.

Coalfield communities "literally fueled the growth of the nation," said Peter Hille, president of the community economic development nonprofit Mountain Association in eastern Kentucky. "There is a debt to be paid. Justice demands we bring new investment to these places: to build a new economy, to revitalize communities and to educate people of all ages to be ready."

Congress continues to consider its role in sustaining coal country, too. In July 2020, U.S. Sen. Tammy Duckworth of Illinois introduced the Marshall Plan for Coal Country Act to establish a plan to stabilize local economies after a mine closure, fund environmental restoration and provide healthcare and higher education for coal workers.

Another, more regional plan titled Reimagine Appalachia, developed by Policy Matters Ohio, West Virginia Center on Budget and Policy, Keystone Research Center, and Kentucky Center for Economic Policy, proposes public investments to expand opportunity, investing in climate-friendly infrastructure and businesses, and promoting workers' rights to rebuild the middle class.

None of these packages come cheap. The geographic scope and sheer depth of the crisis facing coal communities requires a heavy investment that only the federal government is capable of making. Congress has previously intervened for coal miners and their families. In December 2019, lawmakers shifted money from an abandoned mine land fund to prop up the United Mine Workers of America's 1974 retirement plan, which had been squeezed over decades and thrown into uncertainty by the recent bankruptcy of Murray Energy.

Securing support for an expensive, large-scale transition in coal country will be challenging,. On one hand, lawmakers — especially conservative Republicans who represent coal communities on Capitol Hill — may be wary of voting for such a large expenditure after spending trillions on economic stimulus bills amid the coronavirus pandemic. On the other, the transformative effect of the pandemic and a nation in political upheaval may provide a rare opportunity to build political consensus behind a transition package.

"I feel like there's a lot of energy behind systems change right now," said Mary Cromer, deputy director of the Appalachian Citizens Law Center. "There are a lot more discussions about just transition now than there were before. Some of that is because just transition is part and parcel to the rethinking of the way we've ordered our society that's going on right now. I am hopeful about that."

THE FUTURE OF COAL COUNTRY

Regardless of where one stands on climate change, coal's days as the dominant economic force in central Appalachia and Wyoming look to be numbered, even as parts of the industry will likely continue for the foreseeable future.

Appalachia's high-grade, low-sulfur metallurgical coal is poised to grow and contract with global economic cycles even as it contracts over time.

The Powder River Basin remains the cheapest-to-mine thermal coal region in the U.S. with potential to pick up new supply contracts as higher-cost thermal coal mines close elsewhere. Wyoming coal might be the last U.S. thermal coal supplier standing in a permanently shrinking industry.

Those lingering prospects, however, are of little economic consolation to coal communities already in the throes of certain decline.

The future of coal country is still uncertain.

Appalachia has seen downward trends in population and economic metrics for decades, and based on projections, looks unlikely to depart from that path. Wyoming, more prosperous by comparison, faces its biggest economic crisis in history without the prospect of a coal recovery. Without an economic course correction, Wyoming may follow Appalachia into an economic twilight where communities begin transition in patchwork fashion. While some communities attempt to reinvent a future without coal, others hold out hope for coal's rebound, risking hope over a steadily shrinking population and growing poverty amid fraying social institutions and infrastructure.

A possible vision for the long-term future of coal country appears in Fallout 76, a 2018 action role-playing video game set in West Virginia in 2102. Unchecked coal mine fires have turned the southern coalfields into a smog-covered area known as "Ash Heap," while the northern steel-producing region has become "Toxic Valley," replete with industrial waste and polluted waters.

A young miner in Howard B. Lee's book, "Bloodletting in Appalachia," suggested a slightly more positive post-coal scenario in 1956 — an area bereft of anything "except the hoot of the owl and the eerie cry of the whippoorwill," but which would eventually be covered by "beautiful National and State forests."

Neither scenario envisions thriving communities. That vision must come from the people in central Appalachia and Wyoming who want a future for their children and grandchildren.

ACKNOWLEDGEMENTS

Mason Adams would like to thank his family for their patience and support; Dustin Bleizeffer, his collaborator on these stories; Ken Paulman and Katie Klingsporn for their work developing and editing this series; Ivy Brashear, Ariel Fugate and Chris Woolery at the Mountain Association; Kelly Jo Drey; Tom Hansell at Appalachian State University; Carl Shoupe; Clark Derry-Williams and Joshua Macey for their analysis of the Blackjewel bankruptcy; Virginia Del. Terry Kilgore; Beth O'Connor of the Virginia Rural Health Association and Debrin Jenkins of the West Virginia Rural Health Association; Adam Searing at Georgetown University's Center for Children and Families; Teresa Tyson and Paula Hill of the Health Wagon; Pam Thomas at the Kentucky Center for Economic Policy; Mary Cromer at the Appalachian Citizens Law Center; Willie Dodson, Matt Hepler and Cat McCue at Appalachian Voices; Sean O'Leary at the West Virginia Center on Budget and Policy; Lyndsey Gilpin; Dave Mistich; Kate Sheppard; Kim Kelly; Roxy Todd; and fellow journalists telling important stories in Appalachia and rural America.

Dustin Bleizeffer would like to thank his father David Bleizeffer for his 33 years as a coal miner in the Powder River Basin and for his endless guidance and support, my collaboration partner Mason Adams as well as the editorial teams at Energy News Network and WyoFile. Thanks to Mark Haggerty and the rest of the team at Headwaters Economics for critical data and insights, Dr. Robert Godby at the University of Wyoming, including the Haub School of Environment and Natural Resources and the UW Economics Department, Jill Morrison and Shannon Anderson and the rest of the Powder River Basin Resource Council for their decades of research, watchdogging and answering an endless stream of my questions.